ICME-13 Topical Surveys

Series editor

Gabriele Kaiser, Faculty of Education, University of Hamburg, Hamburg, Germany

More information about this series at http://www.springer.com/series/14352

Norma Presmeg · Luis Radford
Wolff-Michael Roth · Gert Kadunz

Semiotics in Mathematics Education

 **Springer** Open

Norma Presmeg
Department of Mathematics
Illinois State University
Normal, IL
USA

Luis Radford
École des Sciences de l'Education
Université Laurentienne
Sudbury, ON
Canada

Wolff-Michael Roth
Lansdowne Professor of Applied Cognitive
  Science
University of Victoria
Victoria, BC
Canada

Gert Kadunz
Department of Mathematics
Alpen-Adria Universitaet Klagenfurt
Klagenfurt
Austria

ISSN 2366-5947   ISSN 2366-5955   (electronic)
ICME-13 Topical Surveys
ISBN 978-3-319-31369-6   ISBN 978-3-319-31370-2   (eBook)
DOI 10.1007/978-3-319-31370-2

Library of Congress Control Number: 2016935590

Printed on acid-free paper

This Springer imprint is published by Springer Nature
The registered company is Springer International Publishing AG Switzerland

Main Topics

- Nature of semiotics and its significance for mathematics education;
- Influential theories of semiotics;
- Applications of semiotics in mathematics education;
- Various types of signs in mathematics education;
- Other dimensions of semiotics in mathematics education.

Contents

Chapter 1
Introduction: What Is Semiotics and Why Is It Important for Mathematics Education?

Over the last three decades, semiotics has gained the attention of researchers interested in furthering the understanding of processes involved in the learning and teaching of mathematics (see, e.g., Anderson et al. 2003; Sáenz-Ludlow and Presmeg 2006; Radford 2013a; Radford et al. 2008, 2011; Sáenz-Ludlow and Kadunz 2016). Semiotics has long been a topic of relevance in connection with language (e.g., Saussure 1959; Vygotsky 1997). But what is semiotics, and why is it significant for mathematics education?

Semiosis is "a term originally used by Charles S. Peirce to designate any sign action or sign process: in general, the activity of a sign" (Colapietro 1993, p. 178). A sign is "something that stands for something else" (p. 179); it is one segmentation of the material continuum in relation to another segmentation (Eco 1986). *Semiotics*, then, is "the study or doctrine of signs" (Colapietro 1993, p. 179). Sometimes designated "semeiotic" (e.g., by Peirce), semiotics is a general theory of signs or, as Eco (1988) suggests, a theory of how signs signify, that is, a theory of sign-ification.

The study of signs has long and rich history. However, as a self-conscious and distinct branch of inquiry, semiotics is a contemporary field originally flowing from two independent research traditions: those of C.S. Peirce, the American philosopher who originated pragmatism, and F. de Saussure, a Swiss linguist generally recognized as the founder of contemporary linguistics and the major inspiration for structuralism. In addition to these two research traditions, several others implicate semiotics either directly or implicitly: these include *semiotic mediation* (the "early" Vygotsky 1978), *social semiotic* (Halliday 1978), various theories of representation (Goldin and Janvier 1998; Vergnaud 1985; Font et al. 2013), relationships amongst sign systems (Duval 1995), and more recently, theories of embodiment that include gestures and the body as a mode of signification (Bautista and Roth 2012; de Freitas and Sinclair 2013; Radford 2009, 2014a; Roth 2010). Components of some of these theories are elaborated in what follows.

The significance of semiosis for mathematics education lies in the use of signs; this use is ubiquitous in every branch of mathematics. It could not be otherwise: the

© The Author(s) 2016
N. Presmeg, *Semiotics in Mathematics Education*,
ICME-13 Topical Surveys, DOI 10.1007/978-3-319-31370-2_1

objects of mathematics are ideal, general in nature, and to represent them—to others and to oneself—and to work with them, it is necessary to employ sign vehicles,[1] which are not the mathematical objects themselves but stand for them in some way. An elementary example is a drawing of a triangle—which is always a particular case—but which may be used to stand for triangles in general (Radford 2006a). As a text on the origin of (Euclidean) geometry suggests, the mathematical concepts are the result of the continuing refinement of physical objects Greek craftsmen were able to produce (Husserl 1939). For example, craftsmen were producing rolling things called in Greek *kulindros* (roller), which led to the mathematical notion of the cylinder, a limit object that does not bear any of the imperfections that a material object will have. Children's real problems are in moving from the material things they use in their mathematic classes to the mathematical things (Roth 2011). This principle of "seeing an A as a B" (Otte 2006; Wartofsky 1968) is by no means straightforward and directly affects the learning processes of mathematics at all levels (Presmeg 1992, 2006a; Radford 2002a). Thus semiotics, in several traditional frameworks, has the potential to serve as a powerful theoretical lens in investigating diverse topics in mathematics education research.

1.1 The Role of Visualization in Semiosis

The sign vehicles that are used in mathematics and its teaching and learning are often of a visual nature (Presmeg 1985, 2014). The significance of semiosis for mathematics education can also be seen in the growing interest of the use of images within cultural science. It was Thomas Mitchel's dictum that the linguistic turn is followed now by a "pictorial turn" or an "iconic turn" (Boehm 1994). The concentration on visualisation in cultural sciences is based on their interest in the field of visual arts and it is still increasing (Bachmann-Medick 2009). But more interesting for our view on visualisation are developments within science which have introduced very sophisticated methods for constructing new images. For example, medical imaging allows us to see what formerly was invisible. Other examples could be modern telescopes, which allow us to see nearly infinite distant objects, or microscopes, which bring the infinitely small to our eyes. With the help of these machines such tiny structures become visible and with this kind of visibility they became a part of the scientific debate. As long as these structures were not visible we could only speculate about them; now we can debate about them and about their existence. We can say that their ontological status has changed. In this regard images became a major factor within epistemology. Such new developments, which

---

[1]A note on terminology: The term "sign vehicle" is used here to designate the signifier, when the object is the signified. Peirce sometimes used the word "sign" to designate his whole triad, object [signified]-representamen [signifier]-interpretant; but sometimes Peirce used the word "sign" in designating the representamen only. To avoid confusion, "sign vehicle" is used for the representamen/signifier.

can only be hinted at here, caused substantial endeavour within cultural science into investigating the use of images from many different perspectives (see, e.g., Mitchell 1987; Arnheim 1969; Hessler and Mersch 2009). The introduction to "Logik des Bildlichen" (Hessler and Mersch 2009), which we can translate as "The Logic of the Pictorial", focusses on the meaning of visual thinking. In this book, they formulate several relevant questions on visualisation which could/should be answered by a science of images. Among these questions we read: epistemology and images, the order of demonstrating or how to make thinking visible.

Let's take a further look at a few examples of relevant literature from cultural science concentrating on the "visual." In their book *The culture of diagram* (Bender and Marrinan 2010) the authors investigate the interplay between words, pictures, and formulas with the result that diagrams appear to be valuable tools to understand this interplay. They show in detail the role of diagrams as means to construct knowledge and interpret data and equations. The anthology *The visual culture reader* (Mirzoeff 2002) presents in its theory chapter "Plug-in theory," the work of several researchers well known for their texts on semiotics, including Jaques Lacan and Roland Barthes, with their respective texts "What is a picture?" (p. 126) and "Rhetoric of the image" (p. 135). Another relevant anthology, *Visual communication and culture, images in action* (Finn 2012) devotes the fourth chapter to questions which concentrate on maps, charts and diagrams. And again theoretical approaches from semiotics are used to interpret empirical data: In "Powell's point: Denial and deception at the UN," Finn makes extensive use of semiotic theories. Even in the theory of organizations, semiotics is used as means for structuring: In his book on *Visual culture in organizations* Styhre (2010) presents semiotics as one of his main theoretical formulations.

1.2 Purpose of the Topical Survey on *Semiotics in Mathematics Education*

Resonating with the importance of semiotics in the foregoing areas, the purpose of this Topical Survey is to explore the significance—for research and practice—of semiotics for understanding issues in the teaching and learning of mathematics at all levels. The structure of the next section is as follows. There are four broad overlapping subheadings:

(1) *A summary of influential semiotic theories and applications*;
(2) *Further applications of semiotics in mathematics education*;
(3) *The significance of various types of signs in mathematics education*;
(4) *Other dimensions of semiotics in mathematics education.*

Within each of these sections, perspectives and issues that have been the focus of research in mathematics education are presented, to give an introduction to what has already been accomplished in this field, and to open thought to the potential for further developments. This Survey is thus an introduction, which cannot be fully comprehensive, and interested readers are encouraged to read original papers cited, for greater depth and detail.

Chapter 2
Semiotics in Theory and Practice in Mathematics Education

2.1 A Summary of Influential Semiotic Theories and Applications

Both Peirce and de Saussure developed theories dealing with signs and significa-tion. Because these differ in a significant aspect—a three-fold relation in the case of the former, a two-fold relation in the case of the latter—Peirce's version goes under *semiotics*, whereas de Saussure's version often is referred to as *semiology*.

2.1.1 Saussure

The basic ideas of this semiotic theory are as follows. Ferdinand de Saussure's (1959) *semiology* was developed in the context of his structural theory of general linguistics. In this theory, a linguistic sign is the result of coupling two elements, a *concept* and an *acoustic image*. To anticipate ambiguities de Saussure proposed to understand the sign as the relation of a signified and a signifier, in a close, insep-arable relationship (metaphorically, like the two sides of a single piece of paper, as he suggests). He uses two now classical diagrams to exemplify the sign. In the first, the Latin word *arbor* [tree] (on the bottom) and the French «arbre» [tree] (on top) form a sign, where the former is the signifier and the latter the signified. In the second diagram, *arbor* is retained as the signifier but the drawing of a tree takes the place of the signified. It is noteworthy that *both* components in this dyad are *psychological*[1]: the acoustic image is a psychological pattern of a sound, which could be a word, a phrase, or even an intonation. These signifiers are arbitrary, in

---

[1]De Saussure uses the French *psychique* [psychical] rather than *mental*, just as Vygotsky will use *psixičeskij* [psychical] rather than *duxovnyj* [mental]. In both instances, the adjective psychological is the better choice because it allows for bodily knowing that is not mental in kind (e.g., Roth 2016b).

© The Author(s) 2016
N. Presmeg, *Semiotics in Mathematics Education*,
ICME-13 Topical Surveys, DOI 10.1007/978-3-319-31370-2_2

the sense that there is no logical necessity underlying them—which accounts for humanity's many languages—but they are not the product of whim because they are socially determined.

This theory has *applications in mathematics education*. Saussure's ideas were brought to the attention of the mathematics education community in the 1990s in a keynote presentation by Whitson (1994), and by Kirshner and Whitson in the context of a book on situated cognition, in a chapter titled "Cognition as a semiosic process: From situated mediation to critical reflective transcendence" (Whitson 1997). Whitson pointed out that for Saussure, although there was interplay between the signified and signifier (denoted by arrows in both directions in his diagrams), the signified, as the top element of the dyad, appeared to dominate the signifier. Lacan (1966) had inverted this relationship, placing the signifier on top of the signified, creating a chain of signifiers that never really attain the signified. This version of semiology was used by Walkerdine (1988), and also became important in Presmeg's research in the 1990s using chains of signification to connect cultural practices of students, in a series of steps, with the canonical mathematical ideas from the syllabuses used by teachers of classroom mathematics (Presmeg 1997). The Lacanian version also is central to a recent conceptualization of subjectivity in mathematics education, which emphasizes that "the signifier does not mark a thing" but "marks a point of pure difference or movement in a discursive chain" (Brown 2011, p. 112). This movement from signifier to signifier creates an effect similar to the interpretant in Peircean semiotics, where one sign–referent relation replaces another sign–referent relation leading to infinite (unlimited) semiosis (Nöth 1990).

The theoretical ideas of de Saussure have not been used as extensively in mathematics education research as those of Peirce, and of Vygotsky (in his earlier notion of *semiotic mediation*), but there are aspects of Saussure's theory that are highly significant. As Fried (2007, 2008) points out, de Saussure's notions of *synchronicity* and *diachronicity* are particularly useful in clarifying ways of looking at both the history of mathematics, and the processes involved in teaching and learning mathematics. The synchronic view is a snapshot in time, while a diachronic analysis is a longitudinal one. A useful botanical metaphor is that synchrony refers to a cross-section of a plant stem, while diachrony takes a longitudinal section. These views are complementary, and both are necessary for a full understanding of a phenomenon (Fried 2007). In mathematics education we are interested not only in understanding *what* is taught and learned in a given situation (synchrony), but particularly in how ideas change—in the *processes* involved as students engage over time with mathematical objects (diachrony). In both the synchronic and diachronic views, sign vehicles play a significant role in standing for mathematical objects; hence both of these distinct viewpoints are useful in semiotic analyses.

The dyadic model of Saussure proved inadequate to account for the results of Presmeg's research, and was later replaced by a Peircean nested model that invoked the interpretant (Presmeg 1998, 2006b).

2.1.2 Peirce

The basic ideas of this theory are as follows. According to Peirce (1992), *trichotomic* is the art of making three-fold divisions. By his own admission, he showed a proclivity for the number three in his philosophical thinking. "But it will be asked, why stop at three?" he wrote (Peirce 1992, p. 251), and his reply to the question is as follows:

> [W]hile it is impossible to form a genuine three by any modification of the pair, without introducing something of a different nature from the unit and the pair, four, five, and every higher number can be formed by mere complications of threes. (p. 251)

Accordingly, he used triads not only in his semiotic model including *object*, *representamen* [sign vehicle], which stands for the object in some way, and *interpretant*, but also in the types of each of these components. These types are not inherent in the signs themselves, but depend on the interpretations of their constituent relationships between sign vehicles and objects. In a letter to Lady Welby on December 23, 1908, he wrote as follows.

> I define a Sign as anything which is so determined by something else, called its Object, and so determines an effect upon a person, which effect I call its Interpretant, that the latter is thereby mediately determined by the former. My insertion of "upon a person" is a sop to Cerberus, because I despair of making my own broader conception understood. I recognize three Universes, which are distinguished by three Modalities of Being. (Peirce 1998, p. 478)

It follows that different individuals may construct different interpretants from the same sign vehicle, thus effectively creating different signs for the same object.

Peirce developed several typologies of signs. Maybe the best known typology is the one based on the kind of relationship between a sign vehicle and its object. The relationship leads to three kinds of signs: iconic, indexical, and symbolic. To illustrate the differences among iconic, indexical, and symbolic signs, it may be useful to look at some of Peirce's examples. In an iconic sign, the sign vehicle and the object share a physical *resemblance*, e.g., a photograph of a person representing the actual person. Signs are indexical if there is some *physical connection* between sign vehicle and object, e.g., smoke invoking the interpretation that there is fire, or a sign-post pointing to a road. The nature of symbolic signs is that there is an element of *convention* in relating a particular sign vehicle to its object (e.g., algebraic symbolism). These distinctions in mathematical signs are complicated by the fact that three different people may categorize the 'same' relationship between a sign vehicle and its object in such a way that it is iconic, indexical, or symbolic respectively, according to their interpretations. In practice the distinctions are subtle because they depend on the interpretations of the learner—and therefore, viewed in this way, the distinctions may be useful to a researcher or teacher for the purpose of identifying the subtlety of a learner's mathematical conceptions if differences in interpretation are taken into account.

Peirce also introduced three conceptual categories that he termed *firstness, secondness,* and *thirdness.* Firstness has to do with that which makes possible the recognizance of something as it appears in the phenomenological realm. It has to do with the *qualia* of the thing. We become aware of things because we are able to recognize their own quale. A quale is the distinctive mark of something, regardless of something else (it is its suchness). "Each quale is in itself what it is for itself, without reference to any other" (Peirce CP 6.224). Thus, what allows us to perceive a red rose is the quality of redness. Were we to be left without qualia, we would not be able to perceive anything. However, quale is not perception yet. It is its mere possibility: it is firstness—the first category of being in Peirce's account. "The mode of being a redness, before anything in the universe was yet red, was nevertheless a positive qualitative possibility" (CP 1.25). Qualia—such as bitter, tedious, hard, heartrending, noble (CP. 1.418)—account hence for the possibility of experience, making it possible to note that something is there, positioned, as it were, in the boundaries of consciousness (Radford 2008a).

Now, the very eruption of the object into our field of perception marks the *indexical* moment of consciousness. It is a moment of actuality or occurrence. Here, we enter *secondness*:

> We find secondness in occurrence, because an occurrence is something whose existence consists in our knocking up against it. A hard fact is of the same sort; that is to say, it is something which is there, and which I cannot think away, but am forced to acknowledge as an object or second beside myself, the subject or number one, and which forms material for the exercise of my will. (Peirce CP 1.358)

Because we have reached awareness, the object now becomes an object of knowledge. But knowledge is not an array of isolated facts or events. Rather, it results from a linkage between facts, and this link, Peirce argues, requires us to enter into a level that goes beyond quality (*firstness*) and factuality (*secondness*). This new level (*thirdness*) requires the use of symbols. Commenting on the subtleties of the interrelationships amongst *firstness, secondness,* and *thirdness* as either ontological or as phenomenological categories Sáens-Ludlow and Kadunz (2016) mention the following:

> Peirce's semiotics is founded on his three connected categories, which can be differentiated from each other, and which cannot be reduced to one another. Peirce argued that there are three and only three categories: 'He claims that he has look[ed] long and hard to disprove his doctrine of three categories but that he has never found anything to contradict it, and he extends to everyone the invitation to do the same' (de Waal 2013, p. 44). The existence of these three categories has been called Peirce's theorem.… He considers these categories to be both ontological and phenomenological; the former deals with the nature of being and the latter with the phenomenon of conscious experience. (Sáenz-Ludlow and Kadunz 2016, p. 4)

Peirce's model includes the need for expression or communication: "Expression is a kind of representation or signification. A sign is a third mediating between the mind addressed and the object represented" (Peirce 1992, p. 281). In an act of communication, then—as in teaching—there are three kinds of interpretant, as follows:

- the "*Intensional* Interpretant, which is a determination of the mind of the utterer";
- the "*Effectual* Interpretant, which is a determination of the mind of the interpreter"; and
- the "*Communicational* Interpretant, or say the *Cominterpretant*, which is a determination of that mind into which the minds of utterer and interpreter have to be fused in order that any communication should take place." (Peirce 1998, p. 478, his emphasis)

It is the latter fused mind that Peirce designated the *commens*. The commens proved to be an illuminating lens in examining the history of geometry (Presmeg 2003).

The complexity and subtlety of Peirce's notions result in opportunities for their use in a wide variety of research studies in mathematics education.

Applications in mathematics education are as follows.
As an example, let us examine the quadratic formula in terms of the triad of iconic, indexical, and symbolic sign vehicles. The roots of the equation $ax^2 + bx + c = 0$ are given by the well-known formula

$$x_{1,2} = \frac{-b \pm \sqrt{b^2 - 4ac}}{2a}.$$

Because symbols are used, the interpreted relationship of this inscription with its mathematical object may be characterized as *symbolic*, involving convention. However, depending on the way the inscription is interpreted, the sign could also be characterized as iconic or indexical. The formula involves spatial shape. In Presmeg's (1985) original research study of visualization in high school mathematics, many of the 54 students interviewed reported spontaneously that they remembered this formula by an image of its shape, an *iconic* property. However, the formula is also commonly interpreted as a pointer (cf. a direction sign on a road): it is a directive to perform the action of substituting values for the variables a, b, and c in order to solve the equation. In this sense the formula is *indexical*. Thus whether the sign vehicle of the formula is classified as iconic, indexical, or symbolic depends on the interpretant of the sign. The phenomenological classification is of importance.

The Peircean approach also was central to a study of how professionals, scientists and technicians, read graphs (Roth and Bowen 2001). In that study, certain aspects of graphs (e.g., the value of a function or its slope at a certain value of the abscissa) were taken as a sign that referred to some biological phenomenon, such as changes in population size. Importantly, the study pointed out that the signs did not just exist. Instead, these needed to emerge from the interpretive activity before they could be related to a biological phenomenon. The results may be understood in terms of the definition of the sign as relation between two segmentations of the material continuum (Eco 1986). As a study of the transformations within a scientific research group shows, not the *material* matters to signification but the *form* of this material (Latour 1993). In the case of familiar signs that appear in familiar circumstances, interpretation is not observed; instead, in reading, users see right through the sign as if it were transparent thereby giving access to the phenomenon itself (Roth 2003a; Roth and Bowen 2003; Roth et al. 2002).

2.1.3 Vygotsky

Basic ideas
Vygotsky's writings spanned a short period of time (from 1915 to 1934). During this period, Vygotsky tackled different problems (creative thinking, special education, cognitive functions, cultural child development, emotions, etc.) from different angles. Contemporary Vygotskian scholars suggest a rough division of Vygotsky's work in terms of *domains* and *moments*. Taking a critical stance towards the current chronology of Vygotsky's works, in his article "The Vygotsky that we (do not) know," Yasnitsky (2011) identifies three main interrelated domains of research that occupied the "Vygotsky circle" (the circle of Vygotsky and his collaborators):

(a) clinical and special education studies;
(b) philological studies (covering problems of language, thinking, and culture); and
(c) studies around affect, will, and action.

González Rey (2011a) suggests an approach to the understanding of Vygotsky's work in terms of three moments, each one marking different emphases that cannot be attributed to a premeditated clear intention:

> Differing emphases that characterize moments in Vygotsky's work did not come about purely as a result of clear intentions. Those moments were also influenced by the effects of the turbulent epoch during which his writings were brought to life, during which the world saw the succession of the Russian Revolution, the First World War, and the rise of Stalin to the top of Soviet political leadership. (González Rey 2011b, p. 258)

The *first moment* covers approximately from 1915 to 1928. Vygotsky's focus here is on the active character of the mind, emotions and phantasy. The main work of Vygotsky's first moment is his 1925 book *The psychology of art* (Vygotsky 1971).

> The central subject of the book suggests a psychology oriented to essential human questions, irreducible to behavior or to an objectivistic view of human beings … in *Psychology of Art,* the basis was created for a psychology capable of studying the human person in all her complexity, as an individual whose psychical processes have a cultural-historical genesis. (González Rey 2011b, p. 259)

The *second moment* goes roughly from 1927 to 1931. It is in the second moment that we find Vygotsky elaborating his concept of sign. Vygotsky's concept of sign was influenced by his work on special education (Vygotsky 1993). In a paper from

1929 he stated that "From a pedagogical point of view, a blind or deaf child may, in principle, be equated with a normal child, but the deaf or blind child achieves the goals of a normal child by different means and by a different path" (p. 60). The special child may achieve her goal in interaction with other individuals. "Left to himself [*sic*] and to his own natural development, a deaf-mute child will never learn speech, and a blind person will never master writing. In this case education comes to the rescue" (p. 168). And how does education do it? Vygotsky's answer is: by "creating artificial, cultural techniques, that is, a special system of cultural signs and symbols" (p. 168). In other words, auxiliary material cultural means (e.g., Braille dots) compensate for differences in the child's sensorial organization. Vygotsky thought of these compensating means as signs.

As a result, in Vygotsky's account, signs are not characterized by their representational nature. Signs are rather characterized by their *functional* role: as external or material means of regulation and self-control. Signs serve to fulfill psychological operations (Radford and Sabena 2015). Thus, in a paper read at the Institute of Scientific Pedagogy at Moscow State University on April 28, 1928, Vygotsky (1993) argued that "A child learns to use certain signs functionally as a means to fulfilling some psychological operation or other. Thus, elementary and primitive forms of behavior become mediated cultural acts and processes" (p. 296). It is from here that Vygotsky developed the idea of the sign both as a *psychological tool* and as a *cultural mediator*.

This two-fold idea of signs allowed him to account for the nature of what he termed the higher psychological functions (which include memory and perception) and to tackle the question of child development from a cultural viewpoint. "The inclusion in any process of a sign," he noted, "remodels the whole structure of psychological operations just as the inclusion of a tool remodels the whole structure of a labor operation" (Vygotsky 1929, p. 421). Signs, hence, are not merely aids to carry out a task or to solve a problem. By becoming included in the children's activities, they *alter* the way children come to know about the world and about themselves. However, the manner in which signs alter the human mind is not related to signs *qua* signs. The transformation of the human mind that signs effectuate is related to their social-cultural-historical role. That is, it depends on how signs *signify* and are used collectively in society. This is the idea behind Vygotsky's famous genetic law of cultural development, which he presented as follows: "Every [psychic] function in the child's cultural development appears twice: first, on the social level, and later, on the individual level" (Vygotsky 1978, p. 57).

Commenting on this idea, Vygotsky (1997) offered the example of language:

> When we studied the processes of the higher functions in children we came to the following staggering conclusion: each higher form of behavior enters the scene twice in its development—first as a collective form of behavior, as an inter-psychological function, then as an intra-psychological function, as a certain way of behaving. We do not notice this fact, because it is too commonplace and we are therefore blind to it. The most striking example is speech. Speech is at first a means of contact between the child and the surrounding people, but when the child begins to speak to himself, this can be regarded as the transference of a collective form of behavior into the practice of personal behavior. (p. 95)

To account for the process that leads from a collective form of behavior to an intra-psychological function Vygotsky introduced the concept of *internalization*. He wrote: "We call the internal reconstruction of an external operation internalization" (Vygotsky 1978, p. 56). To illustrate the idea of internalization Vygotsky (1978) provided the example of pointing gestures:

> A good example of this process may be found in the development of pointing. Initially, this gesture is nothing more than an unsuccessful attempt to grasp something, a movement aimed at a certain object which designates forthcoming activity. The child attempts to grasp an object placed beyond his reach; his hands, stretched toward that object, remain poised in the air. His fingers make grasping movements. At this initial stage pointing is represented by the child's movement, which seems to be pointing to an object—that and nothing more. When the mother comes to the child's aid and realizes his movement indicates something, the situation changes fundamentally. Pointing becomes a gesture for others. The child's unsuccessful attempt engenders a reaction not from the object he seeks but from another person. Consequently, the primary meaning of that unsuccessful grasping movement is established by others. Only later, when the child can link his unsuccessful grasping movement to the objective situation as a whole, does he begin to understand this movement as pointing. (p. 56)

To sum up, in the second moment of Vygotsky' work there is a shift from imagination, phantasy, emotions, personality, and problems of personal experience to an instrumental investigation of higher psychological functions. This instrumental investigation revolved around the notion of signs as a tool and the concomitant idea of *semiotic mediation.*

González Rey (2009) qualifies this moment as an instrumentalist "objectivist turn," that is, a turn in which the subjective dimension that was at the heart of Vygotsky's first moment shades away to yield room to the study of "internalization of prior external processes and operations" (p. 63). He continues:

> Vygotsky explained the transition from intermental to intra-mental, a specifically psychical field, through internalization, which still represents a very objectivistic approach to the comprehension of the psyche. This comprehension of that process does not lend a generative character to the mind as a system, recognizing it only as an internal expression of a formerly inter-mental process. Several Soviet psychologists also criticized the concept of internalization in different periods. (p. 64)

In the *third moment* (roughly located during the period from 1932 to 1934), Vygotsky returned with new vigor to some ideas of the first moment, such as the unity between cognition and emotion, and the interrelationship of social context and subjective experience. During those years, he pointed out that the genetic origin of all higher psychological functions *was* a soci(et)al relation (Vygotsky 1989). He did not write that there was something *in* the relation that then was transferred mysteriously into the person. Instead, the soci(et)al relation itself *is* the higher function. That is, the developing individual already contributes to the realization of the higher function; it is when s/he assumes all parts of the relation that the higher function can be ascribed to the individual (e.g., Roth 2016b). Moving away from the mechanist or instrumental turn of the second period, questions of the generative power of the mind that we find in his study of Hamlet came to the fore again (Vygotsky 1971).

Although the aforementioned moments are relevant in the understanding of Vygotsky's ideas and in particular the understanding of Vygotsky's semiotics, we should not think that the problems that Vygotsky tackled were marked differently from one moment to the other. These moments may be understood in terms of *focus*. We should not think for instance that, in the second period, signs are strictly thought of as mediators per se; they were associated with *meaning* too. Already in his work on special education Vygotsky (1993) noted that "Meaning is what is important, not the signs in themselves. We may change the signs but the meaning will be preserved" (p. 85). The problem of meaning is tackled again in his later work, this time in the context of a communicative field that is common to the participants in a relation (Roth 2016a). In some notes from an internal seminar in 1933—hence a short time before Vygotsky' death—a seminar in which Vygotsky (1997) summarized his group's accomplishments and new research avenues, we read: "the problem of meaning was already present in [our] older investigations. Whereas before our task was to demonstrate what 'the knot' and logical memory have in common, now our task is to demonstrate the difference that exists between them" (pp. 130–131).

Some Russian scholars in the cultural-historical tradition now suggest that towards the very end of his life, Vygotsky was moving away from the idea of sign mediation, developing instead the idea of a semiotic or intersubjective speech field (e.g., El'konin 1994; Mikhailov 2006). One indication of this move is noticeable at the very end of the posthumously published *Thinking and speech* (Vygotsky 1987), where he notes that the word is impossible for an individual, but is a reality for two. Even if a person writes into a diary, s/he still is relating to herself as to another. Thus, signs generally and language specifically—generally theorized as the mediators between subject and material world or between two subjects—"are given to the child *not as an ensemble of mediators* between the child and nature, but, in fact, as subjectively his *own*; for all of these things are subjectively 'everyone's'" (Mikhailov 2001, p. 27, original emphasis, underline added). This insight implies that intersubjectivity is not problematic, as often assumed; it is a modality of the semiotic speech field. Instead, subjectivity is the result of participation in relations with others, relations that take place in a semiotic field. The very notion of a mediator is the result of, or gives rise to, the Cartesian division between body and mind or psychic-physical parallelism (Mikhailov 2004). To overcome the dangers of the split between body and mind, Vygotsky was turning to a Spinozist idea, where material bodies and culture (mind) are but two (contradictory) manifestations of *one substance*. Based on the idea of inner contradictions, Marxist psychologists have shown a possible evolutionary and cultural-historical trajectory that led from the first cell to the human psyche of today, including its languages and tools (Holzkamp 1983; Leontyev 1981).

Applications to mathematics education
Vygotsky's work has inspired mathematics education researchers interested in the question of teaching and learning. Arzarello and his collaborators have been interested in the *evolution of signs*. To do so, they have developed the theoretical

construct of *semiotic bundle* (Arzarello 2006; Arzarello et al. 2009). This notion encompasses signs and semiotic systems such as the contemporary mathematics sign systems of algebra, Cartesian graphs, but also gestures, writing, speaking, and drawing systems. Arzarello et al.'s work is located within a broader context of multimodality that they explain as coming from neuroscience studies that have highlighted the role of the brain's sensory-motor system in conceptual knowledge, and also from communication and multiple modes to communicate and to express meanings. Within this perspective, a semiotic bundle is defined as

> *a system of signs* […] that is produced by one or more interacting subjects and that evolves in time. Typically, a semiotic bundle is made of the signs that are produced by a student or by a group of students while solving a problem and/or discussing a mathematical question. Possibly the teacher too participates to this production and so the semiotic bundle may include also the signs produced by the teacher. (Arzarello et al. 2009, p. 100)

Paying attention to a wide variety of means of expression, from the standard algebraic or other mathematical symbols to the embodied ones, like gestures and gazes, and considering them as semiotic resources in teaching and learning processes, the concept of semiotic bundle goes beyond the range of semiotic resources that are traditionally discussed in mathematics education literature (e.g., Duval 2006; Ernest 2006). Arzarello and collaborators track the students' learning through the evolution of signs in semiotic bundles.

Bartolini Bussi and Mariotti (2008) have focused on the concept of semiotic mediation, in particular in the case of artifacts and signs. In their seminal paper they distinguished between mediation and semiotic mediation. Mediation involves four terms—someone who mediates (the mediator); something that is mediated; someone or something subjected to the mediation (the mediatee), and the circumstances for mediation (Hasan 2002). Semiotic mediation, in Bartolini Bussi and Mariotti's (2008) account, appears as a particular case of mediation:

> Within the social use of artifacts in the accomplishment of a task (that involves both the mediator and the mediatees) shared signs are generated. On the one hand, these signs are related to the accomplishment of the task, in particular related to the artifact used, and, on the other hand, they may be related to the content that is to be mediated … Hence, the link between artifacts and signs overcomes the pure analogy in their functioning in mediating human action. It rests on the truly recognizable relationship between particular artifacts and particular signs (or system of signs) directly originated by them. (p. 752)

Within this context, "any artifact will be referred to as tool of semiotic mediation as long as it is (or it is conceived to be) intentionally used by the teacher to mediate a mathematical content through a designed didactical intervention" (p. 754).

Anna Sfard (2008) has also drawn on Vygotsky in her research on thinking, which she conceives of as the individualized form of interpersonal communication. She wrote as follows:

> Human communication is special, and not just because of its being mainly linguistic—the feature that, in animals, seems to be extremely rare, if not lacking altogether. It is the role communication plays in human life that seems unique. The ability to coordinate our activities by means of interpersonal communication is the basis for our being social

creatures. And because communication is the glue that holds human collectives together, even our ability to stay alive is a function of our communicational capacity. We communicate in order to coordinate our actions and ascertain the kind of mutuality that provides us with what we need and cannot attain single-handedly. (p. 81)

From this viewpoint, she defines thinking as follows: "Thinking is an individualized version of (interpersonal) communicating" (p. 81), that is, "as one's communication with oneself" (Sfard 2001, p. 26). An important role is ascribed to communication mediators, which are "perceptually accessible objects with the help of which the actor performs her prompting action and the re-actor is being prompted" (Sfard 2008, p. 90). They include "artifacts produced specially for the sake of communication (p. 90). Within this context, Sfard conceptualizes learning as changes in discourse. More precisely,

learning mathematics means changing forms of communication. The change may occur in any of the characteristics with the help of which one can tell one discourse from another: words and their use, visual mediators and the ways they are operated upon, routine ways of doing things, and the narratives that are being constructed and labelled as "true" or "correct." (Sfard 2010, p. 217)

In the next section we turn to semiotics in mathematics education.

2.2   Further Applications of Semiotics in Mathematics Education

The summary of influential semiotic theories conducted in the previous section provided an idea of the impact of these theories in mathematics education. In this section we discuss in more detail the impact that semiotics has had in specific problems of mathematics teaching and learning.

As previously mentioned, two different approaches can be distinguished within semiotics, depending on how signs are conceptualized: a representational one, in which signs are essentially representation devices, and one in which signs are conceptualized as mediating tools (Radford 2014b). There is still a third approach—a dialectical materialist one—in which signs and artefacts are a fundamental part of mathematical activity, yet they do not represent knowledge, nor do they mediate it (Radford 2012). This is the approach to signs, artifacts, and material culture in general that is featured in the theory of objectification (Radford 2006b, 2008b, 2013b, 2015a, b). Such a conception of signs and artefacts is consubstantial with the conception of the dialectical materialist idea of *activity*. This conception of activity is very different from usual conceptions that reduce activity to a series of actions that an individual performs in the attainment of his or her goal. The latter line of thinking reduces activity to a functional conception: activity amounts to the deeds and doings of the individuals. Activity in the theory of objectification does not merely mean to do something. Activity (Tätigkeit in German and deyatel'nost' in Russian) refers to a dynamic *system* geared to the satisfaction of collective needs that rests on:

(1)   specific forms of human collaboration; and
(2)   definite forms of material and spiritual production.

Activity as Tätigkeit should not be confounded with activity as Aktivität/ aktivnost', that is, as being simply busy with something (Roth and Radford 2011). Activity as Tätigkeit does not have the utilitarian and selfish stance that it has come to have in capitalist societies. Activity as Tätigkeit is a social form of joint action through which individuals produce their means of subsistence and "comprises notions of self-expression, rational development, and aesthetic enjoyment" (Donham 1999, p. 55). More precisely, it is a form of life. Activity as Tätigkeit is the endless process through which individuals inscribe themselves in society.

To avoid confusions with other meanings, Activity as Tätigkeit is termed *joint labor* in the theory of objectification. The concept of joint labor allows one to revisit classroom teaching and learning activity. It allows one to see teaching and learning activity not as two separate activities, one carried out by the teacher (the teacher's activity) and another one carried out by the student (the student's activity), but as a single and same activity: the same teachers-and-students *joint labor*. The concept of joint labor is central to the theory of objectification: It is, indeed, through joint labor that, in this theory, the students are conceived of as encountering and becoming gradually aware of culturally and historically constituted forms of mathematics thinking. The joint-labor bounded encounters with the historical forms of mathematics thinking are termed *processes of objectification*. The theory of objectification is an attempt to understand learning not as the result of the individual student's deeds (as in individualist accounts of learning) but as a cultural-historical situated processes of knowing and becoming. It seeks to study the manners by which the students become progressively aware of historically and culturally constituted forms of thinking and acting, and how, as subjectivities in the making, teachers and students position themselves in mathematical practices.

The semiotic dimension of the theory of objectification is apparent at different levels:

(1)   The first one is the level of the material culture (signs, artefacts, etc.).
(2)   The second one is a suprastructural level of cultural meanings that shape and organize joint labor.

As mentioned before, signs and artefacts are not considered representational devices or aiding tools. But neither are they considered as the mere stuff that we touch with our hands, hear with our ears, or perceive with our eyes. They are considered as bearers of sedimented human labor. That is, they are bearers of human intelligence and specific historical forms of human production that affect, in a definite way, the manner in which we come to know about the world.

Now, the fact that signs and artifacts are bearers of human intelligence does not mean that such an intelligence is transparent for the student who resorts to them. Leont'ev (1968) notes:

If a catastrophe would happen to our planet so that only small children would survive, the human race would not disappear, but the history of humanity would inevitably be interrupted. The treasures of material culture would continue to exist, but there would be no one who would reveal their use to the young generations. The machines would be idle, the books would not be read, artistic productions would lose their aesthetic function. The history should restart from the beginning. (p. 29)

To fulfil their function and to release the historical intelligence embedded in them, signs and artefacts have to become an integral part of joint labor. In doing so, they become a central part of the processes through which students encounter culturally and historically constituted forms of thinking and acting. All the semiotic resources that students mobilize in order to become aware of such historical forms of thinking and action are termed *semiotic means of objectification* (Radford 2002a, 2003).

The term objectification has its ancestor in the word object, whose origin derives from the Latin verb *obiectare*, meaning "to throw something in the way, to throw before" (Charleton 1996, p. 550). The suffix–*tification* comes from the verb facere meaning "to do" or "to make" (p. 311), so that in its etymology, objectification becomes related to those actions aimed at bringing or throwing something in front of somebody or at making something an object of awareness or consciousness (Radford 2003). Semiotic means of objectification may include material mathematical signs (e.g. alphanumeric formulas and sentences, graphs, etc.) objects, gestures, perceptual activity, written language, speech, the corporeal position of the students and the teacher, rhythm, and so on.

An example of research
In order to show the pragmatic implications of the theoretical ideas presented in the forgoing, the following is one example, in more detail, of research studies in this paradigm.

Radford (2010a) discusses an example of pattern generalization in which Grade 2 seven-to-eight-year-old students were invited to draw Terms 5 and 6 of the sequence shown in Fig. 2.1.

Figure 2.2 shows two paradigmatic answers provided by two students: Carlos and James.

These answers suggest that the students were focusing on *numerosity*. Such a strategy may prove difficult to answer questions about remote terms, such as Terms 12 or 25, which was in fact the case in this classroom. The students worked by themselves more than 30 min. When the teacher came to see the students, she engaged them in an exploration of the patterns in which a spatial structure came to the fore: to see the terms as made up of two rows (see Fig. 2.3).

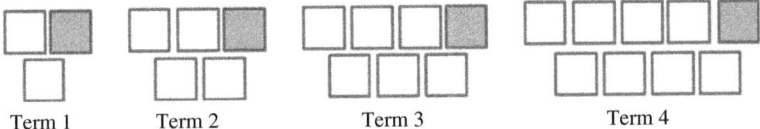

Term 1   Term 2   Term 3   Term 4

Fig. 2.1 The first terms of a sequence that Grade 2 students investigated in an algebra lesson

Fig. 2.2 *Left* Carlos, counting aloud, points sequentially to the squares in the *top row* of Term 3. *Middle* Carlos's drawing of Term 5. *Right* James's drawing of Term 5

Fig. 2.3 *Left* The teacher pointing to the *bottom rows*. *Right* Students and the teacher counting together

The teacher says, "We will just look at the squares that are on the bottom." At the same time, to visually emphasize the object of attention and intention, the teacher makes three consecutive sliding gestures, each one going from the bottom row of Term 1 to the bottom row of Term 4. Figure 2.3, left, shows the beginning of the first sliding gesture. The teacher continues: "Only the ones on the bottom. Not the ones that are on the top. In Term 1 (she points with her two index fingers to the bottom row of Term 1; see Fig. 2.3, left). How many [squares] are there?" Pointing, one of the students answers "one." The teacher and the students continue rhythmically exploring the bottom row of Terms 2, 3, and 4, and also, through gestures and words, the non-perceptually accessible Terms 5, 6, 7, and 8. Then, they turn to the top row.

This short excerpt illustrates some semiotic means of objectification: gestures, words, the mathematical figures, body position, perceptual activity, and rhythm. They are at work in a crucial part of the students' joint labor process: the *process of objectification*, that is, the progressive, sensuous, and material encountering and making sense of a historically and culturally form of thinking mathematically. The theory of objectification is a dialectical materialist theory based on the idea of Otherness or alterity. Learning is to encounter something that is *not me*. The theory of objectification posits the subject and the object as *heterogeneous* entities. In encountering the cultural object, that is to say, an object of history and culture, it *objects* me. Etymologically speaking, it means that I *feel* it as something *alien* and, in the poetic encounter of the object and myself, I come to cognize it, not only cognitively but emotionally, sensuously, even if I do not agree with it. This encounter of the object is what objectification is about (and from whence the theory takes its name).

In the example under consideration, within the teacher-students joint labor occurs a process of objectification in the course of which the students start noticing a culturally and historically constituted theoretical way of seeing and gesturing—seeing and gesturing algebraically. The students start discerning a new way of perceiving out of which an algebraic numerical-spatial structure becomes *apparent* and can be now applied to other terms of the sequence that are not in the students' perceptual field.

In this example, imagination is central in the entailed process of objectification. Imagining the non-perceptually accessible terms is a fully sensuous process out of which an algebraic sense of the functional relations between the number of the terms and the number of squares in their bottom and top rows starts to emerge. However, as the example suggests, the semiotic means of objectification do not operate isolated from each other. On the contrary, they operate through a complex coordination of various sensorial modalities and semiotic registers that the students and teachers mobilize in a process of objectification.

The *segment of joint labor* where such a complex coordination of sensorial modalities and semiotic registers occur is called a *semiotic node* (Radford et al. 2003). In the previous example, when the students are counting along with the teacher, the semiotic node is a segment of joint labor where signs and sensuous modalities cooperate in order for the students to notice, grasp and become aware, or conscious of, an algebraic structure in the terms of the sequence. The segment of joint labor that constitutes the semiotic node includes signs on the activity sheet, the teacher's sequence of gestures, the words that the teacher and the students pronounce simultaneously, the coordinated perception of the teacher and the students, the corporeal position of the students and the teacher, and rhythm as an encompassing sign that links gestures, perception, speech, and symbols. The semiotic node is in this case a collective phenomenon out of which the algebraic structure appears in sensible, collective consciousness.

Semiotic bundles (Arzarello 2006) and semiotic nodes (Radford et al. 2003) are two theoretical constructs that attend to different things. A semiotic bundle is made of the signs produced by teachers and students in problem solving. A semiotic bundle attends to the evolution of signs:

> Looking at the evolution of the students' signs, the teacher can gain clues with respect to the students' understanding: the multimodal aspects of the activity can therefore help her decide whether or not to intervene in order to support the students. (Radford et al. in press)

A semiotic node is not made of signs. It is a *segment* of the teacher's and students' *joint labor* where a complex coordination of sensorial modalities and semiotic registers occurs in a process of knowledge objectification. Thus, while semiotic bundles focus on signs, semiotic nodes focus on joint labor (Activity as

Tätigkeit). It is a segment of joint labor in which a progressive encounter with mathematics occurs. The encounter is described in terms of noticing, grasping, making sense, awareness, becoming conscious. This is why semiotic nodes focus on attention, intention, and meaning making.

Within dialectical materialism philosophy, which informs the theory of objectification, consciousness is a central concept. Vygotsky dealt with the concept of consciousness throughout his academic life, from the famous "Consciousness as a problem in the psychology of behavior" 1925 paper (Vygotsky 1979) up to his last works (e.g., *Thinking and speech*, Vygotsky 1987). If we remove the concept of consciousness (and stop talking about noticing, becoming aware, etc.) the theory of objectification just collapses (as the theory of didactic situations (Brousseau 1997) would do if you removed from there the concept of situation or if you remove the idea of autonomous child from constructivism). The theory of objectification would simply collapse because it defines learning as processes of objectification, and these are a *problem of consciousness*. There are many theories of learning that do not need to refer to consciousness.

Of course, in dialectical materialism, consciousness is not the metaphysical construct of idealism—something buried in the depths of the human soul. From the dialectical materialist perspective adopted in the theory of objectification, consciousness is a concrete theoretical construct. As Vygotsky (1979, p. 31) stated in 1925: "consciousness must be seen as a particular case of the social experience." The structure of consciousness "is the relation [of the individual] with the external world" (Vygotsky 1997, p. 137). Leont'ev (2009) insisted on the idea that consciousness cannot be understood without understanding the individual's activity (Tätigkeit):

> Man's (sic) consciousness… is not additive. It is not a flat surface, nor even a capacity that can be filled with images and processes. Nor is it the connections of its separate elements. It is the internal movement of its "formative elements" geared to the general movement of the activity which effects the real life of the individual in society. Man's activity is the substance of his consciousness. (p. 26).

From this perspective, consciousness is open to empirical investigation, for as Voloshinov (1973, p.11) put it, "consciousness… is filled with signs. Consciousness becomes consciousness only once it has been filled with… (semiotic) content, consequently, only in the process of social interaction". In other words, the fabric of consciousness is semiotic.

Let us come back to the previous example, to a passage where the teacher and the students are exploring the bottom row of the terms—a passage that occurred a bit later than the previous one. The teacher points rhythmically to the terms one after the other and says:

1. Teacher: Now it's Term 8! (The teacher comes back to Term 1. She points again with a two-finger indexical gesture to the bottom row of Term 1) Term1, has how many [squares] on the bottom?
2. Students: 1.

3. Teacher: (Pointing with a two-finger indexical gesture to the bottom row of Term 2) Term 2?
4. Students: 2!
5. Teacher: (Pointing with a two-finger indexical gesture
6. to the bottom row of Term 3) Term 3?
7. Students: 3!
8. Teacher: (Pointing with a two-finger indexical gesture to the hypothetical place where the bottom row of Term 4 would be) Term 4?
9. Students: 4!
10. Teacher: (Pointing as above) Term 5 (see Fig. 2.2)?
11. Students: 5!
12. Teacher: (Pointing as above) Term 6?
13. Students: 6!
14. Teacher: (Pointing as above) Term 7?
15. Students: 7!
16. Teacher: (Pointing as above) Term 8?
17. Students: 8!
18. Sandra: There would be 8 on the bottom!

This segment of the students-and-teacher joint labor is the semiotic node. The excerpt allows us to see the complex coordination of sensorial modalities and semiotic registers that collaborate in the students' awareness or consciousness of a functional relationship between the number of the terms and the number of squares on the top row of the terms (Fig. 2.4).

We can turn now to the second semiotic dimension in the theory of objectification. We mentioned that this dimension has to do with a supra-structural level of cultural meanings that shape and organize joint labor. Joint labor, indeed, is not something that unfolds spontaneously. Although it is unpredictable and cannot be

Fig. 2.4 The teacher pointing the *bottom row* of Term 5

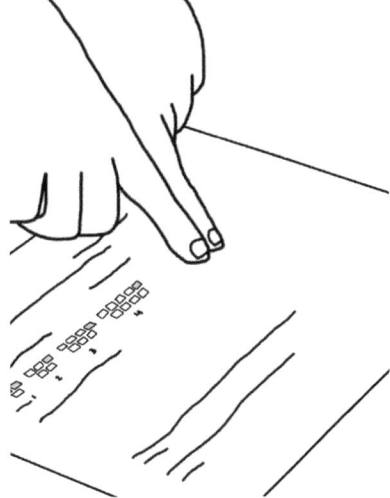

anticipated in all its details, it is shaped culturally. In the previous example, the teacher did not know how the students would engage by themselves in the generalizing tasks, nor did she know how the students would react to her invitation to explore the sequence in terms of rows. Unpredictable as it may be, joint labor is nonetheless shaped by forms of human collaboration and modes of knowledge production that find their meaning in culture and society and that are fostered by the school. These meanings have to do with conceptions about the mathematics to be taught and learned, how it should be taught and learned, and ultimately with our understanding of ourselves as humans. The mathematics to be taught and learned conveys views about the purpose and the nature of mathematic (e.g., the nature of mathematical truth, the relationship between mathematics and the empirical world), the legitimacy of the methods of investigation, etc. Our understanding of ourselves as humans conveys views about the child, and specifically the student, and also about the teacher. What are their roles? The teaching and learning of mathematics differ as we move from one historical period to another. For instance, teaching and learning of mathematics were very different in the Mesopotamian House of Scribes, in Plato's time, in the Renaissance schools of Abacus and today.

What is different is not only the content, but also the conception of mathematics, the conception of the teacher and the idea of the student. The teacher appears as a financial advisor (Radford 2014c). She cannot produce for the student, but can help the student to produce more. Those meanings of mathematical truths, mathematical methods of investigation and production, as well as the meanings of students and teachers, find their legitimacy in cultural meanings that go in general undisputed, although an increasing effort is been made to expose and discuss them (see, e.g., Alrø et al. 2010; Pais 2013; Popkewitz 2004; Skovsmose 2008). In Radford (2006b, 2008b) these meanings are considered to be part of a symbolic suprastructure that is called *Semiotic Systems of Cultural Significations*. Radford and Empey (2007) resort to these systems in order to investigate the historical creation of new cultural forms of mathematical understanding and novel forms of subjectivity. They present two case studies. One devoted to the Western Late Middle Ages and Renaissance and the other to the Buwayhid period of medieval Islam.

2.3 The Significance of Various Types of Signs in Mathematics Education

2.3.1 Embodiment, Gestures, and the Body in Mathematics Education

Embodiment has gained a great deal of attention in the past few years. Some of the theories of embodiment in mathematics education have been influenced by Piaget's genetic epistemology and the Kantian idea of the schema. This is the case of the so-called "process-object" theories; that is, theories that conceive of thinking as moving from the learner's actions to operation knowledge structures. Two examples are APOS theory (Dubinsky 2002; Dubinsky and McDonald 2001) and the "three worlds of mathematics" (Tall 2013). The *first world of mathematics* refers to *conceptual embodiment,* which builds on perception and action to develop mental images that "become perfect mental entities" (Tall 2013, p. 16). For instance, "the number line develops in the embodied world from a physical line drawn with pencil and ruler to a 'perfect' platonic construction that has length but no thickness (Tall 2008, p. 14). The meaning of the term *embodiment* in the "three worlds of mathematics" approach is explained as something that is "consistent with the colloquial notion of 'giving a body' to an abstract idea" (Tall 2004, p. 32). As a result, embodiment remains a general category; the fate of embodied actions is to be superseded by flexible actions with symbols (Radford et al. in press).

Embodiment theories coming from the field of cognitive linguistics generally assume a mediator, such as (bodily) schemas, that relate the mind to the material world (e.g., Johnson 1987; Lakoff 1987). The schemas themselves are the result of bodily engagements with the world that are developed into new concepts by means of metaphorization (Lakoff and Núñez 2000). Thus, for example, one study analyzes the hand/arm gestures that a mathematics professor produces in the course of a lecture on the mathematical concept of continuity (Núñez 2009). In this case, the source-path-goal schema mediates between the mathematician's concept of continuity and the bodily gestural expression. The problem of these embodiment theories is that something else is assumed as *prior* to the movement, something that then is enacted by the body; and this assumption conjures again the specter of Cartesianism (Sheets-Johnstone 2009). Following up on this critique, and grounded in the works of a largely forgotten French philosopher P. Maine de Biran (e.g., 1841) and the uptake of his work in material phenomenology (Henry 2000), an approach to semiotics has been proposed in which bodily movement takes precedence over the schemas or concepts ordinarily taken as that which comes to be enacted in gestures and other signifying body movements and positions (Roth 2012). Using examples from university physics lectures, the study shows how the signs (parts of a graph) are the endpoint and the distillate of movements that generate a field and perceptual structures. The study proposes a model where body movements self-affect so as to lead to a bifurcation in which the sign is born as the relation between two movements.

Before the sign—understood as the relation between two segmentation of matter—can be read (transparently) or interpreted, it actually has to come into being (Roth 2008). From a phenomenological perspective, "*what* is taken to be sign initially has had to be accessible in it and has to be captured *prior* to being made sign" (Heidegger 1977, p. 81). In one study of graphing in a fish hatchery, a fish culturist with high school certification was looking at two distributions, one representing the weight of 100 fish, the other one the lengths of the same fish; she then showed on the graphs which of the fishes were short and fat versus those that were long and skinny (Roth 2016b). Moreover, she related the graphs to a condition coefficient, which is calculated by dividing fish weight by the cube of the length and multiplying the result by 100,000. The study showed that over the years of literally handling and inspecting fishes, and entering their lengths and weights into a spreadsheet that immediately plotted graphs, the fish culturist had developed a feel for both graphs and the fishes.

Longitudinal studies among high school students provided insights about the emergence of signs from work generally and hands-on activities specifically (Roth 2003b; Roth and Lawless 2002). These studies show a progression from hand/arm movements that either did work (i.e., *ergotic* gestures) or found something out by means of sense (i.e., *epistemic* gestures) sometime later, had *symbolic* function. A subsequent study of the emergence and evolution of sign systems (Roth 2015) suggested that signs initially are immanent to the work activity; and in the transition to symbolic function, they transcend the activity. Once there are symbolic functions, these may be replaced by other signs. In the process, motivated signs (i.e., signs involving iconic relations) develop into arbitrary relations.

2.3.2   Linguistic Theories and Their Relevance in Mathematics Education

An important contribution to a theory of signs can be found in the *Philosophical Investigations* (Wittgenstein 1953/1997). In this pragmatic approach to the sign, the focus is on use rather than meaning. In fact, Wittgenstein notes that the "philo-sophical concept of meaning has its place in a primitive idea of the way language functions" (p. 3 [§2]). He illustrates this in articulating how to take the verb "to signify." He suggests marking a tool used in building something with a sign; when the master builder shows the helper another instance of the same sign, the latter will get the tool and bring it to the master. Throughout his book, Wittgenstein uses many cases to exemplify that the use and function of signs matters rather than some metaphysical concept or idea. An application of this in mathematics education shows that signs denoting "concepts" can be taken concretely, referring us to the many concrete ways in which some sign finds appropriate use (Roth in press). A sign, then, is grounded in and indexes all those concrete situations in which it has found some appropriate use. The sign "cylinder," used by a child, then is a placeholder for all the situations in which s/he has encountered and made use of it (e.g., in asking questions, making constative statements, or contesting observation categoricals of others). Here, it is apparent that the usage of signs is tied to concrete situations. This is why Wittgenstein defines a *language-game* as a whole that weaves together a concrete human activity and the language that is part of the work of accomplishing the work. This program is taken up in ethnomethodological research on mathematics, which studies the actual living work of doing mathe-matics and how signs (e.g., those required to prove Gödel's theorem) are mobilized to do work and to formulate the work of doing (Livingston 1986). This non-metaphysical approach to signs is taken up, for example, in studies of ethno-mathematics (e.g., Knijnik 2012; Vilela 2010).

Another important language theory was developed in the circle surrounding the literary critic and philosopher M.M. Bakhtin (e.g., 1981). This theory, often referred to as dialogism, has fundamental commonalities with the approach to language taken by the last works of Vygotsky (e.g., Mikhailov 2001; Radford 2000; Roth 2013), even though some authors appear to be unaware of the fundamentally dialogic approach in Vygotsky (e.g., Barwell 2015). In both theories, there is a primacy of the dialogue as the place where linguistic competence emerges; dialogue with others is the origin of individual speaking and thinking (Vygotsky 1987). But dialogue always presupposes familiarity with the situation and the purpose of speaking. Moreover, dialogue requires the sign (word) to be a reality for all participants (Vygotsky 1987; Vološinov 1930). The sign, as the commodity in a dialectical materialist approach to political economy, is not a unitary thing (Roth 2006, 2014). Instead, the sign is conceived as a phenomenon harboring an *inner* contradiction that manifests itself in the different ways that individuals may use a sign (word). The most important aspect of the dialogical approach is that it inherently is a dynamical conception of language and sign systems generally and of ideas specifically. Thus, in use, signs (language) live; but because they live, signs (language) change at the very moment of their use (Bakhtin 1981; Vološinov 1930). That is, whenever we use signs, whether in dialogue with others or in dialogue with ourselves, signs and the ideas developed with them evolve (Bakhtin 1984). Thus, in this theory *dialogue* does not require two or more persons, and monologue may occur even in the exchange between two persons. For Bakhtin, dialogical speech requires the relation between two voices that build on and transform one another; and therefore such speech is never final. This dialogue may occur within one person, as exemplified in Dostoevsky's novella *Notes from the underground;* On the other hand, the talk involving two individuals may simply be a way of expounding a pre-existing, finalized truth, as exemplified in the works of the late Plato, where the "monologism of the content begins to destroy the form of the Socratic dialogue" (Bakhtin 1984, p. 110). The Bakhtinian approach is found particularly suited in studies of mathematical learning that focus on mathematical learning (e.g., Barwell 2015; Kazak et al. 2015) and in studies of the narrative construction of self and the subject of mathematical activity (e.g., Braathe and Solomon 2015; Solomon 2012).

Another line of inquiry comes from Halliday's social semiotics (Morgan 2006, 2009, 2012). Researching from this perspective Morgan (2006) notes:

> An important contribution of social semiotics is its recognition of the range of functions performed by use of language and other semiotic resources. Every instance of mathematical communication is thus conceived to involve not only signification of mathematical concepts and relationships but also interpersonal meanings, attitudes and beliefs. This allows us to address a wide range of issues of interest to mathematics education and helps us to avoid dealing with cognition in isolation from other aspects of human activity. (p. 220)

Indeed, in this line of inquiry, there is an intention to go beyond the traditional view that reduces individuals to the cognitive realm and that reduces the student to a cognitive subject. "Individuals do not speak or write simply to externalise their personal understandings but to achieve effects in their social world" (Morgan 2006, p. 221). As a result, "Studying language and its use must thus take into account both the immediate situation in which meanings are being exchanged (the context of situation) and the broader culture within which the participants are embedded (the context of culture)" (Morgan 2006, p, 221). The *context of culture* theoretical construct is oriented, like the Semiotic Systems of Cultural Significations in the theory of objectification alluded to before, towards the understanding of classroom practices as loci of production of subjectivities within the parameters of culture and society: "The context of culture includes broader goals, values, history and organizing concepts that the participants hold in common. This formulation of context of culture suggests a uniformity of culture both between and within the participants" (Morgan 2006, p. 221). Yet, as Morgan (2006) argues, such a uniformity is relative and needs to be nuanced in order to account for the variety of responses, behaviors, meaning-making, and language use that are found in individuals of a same culture:

> [T]he notion of participation in multiple discourses will be used as an alternative way of conceptualising this level of context. Importantly, however, the thinking and meaning making of individuals is not simply set within a social context but actually arises through social involvement in exchanging meanings. (p. 221)

2.4 Other Dimensions of Semiotics in Mathematics Education

In this section we discuss briefly three interesting questions that have been the object of scrutiny in semiotics and mathematics education research. The first one is the relationship among sign systems (e.g., natural language, diagrams, pictorial and alphanumeric systems) and the translation between sign systems in mathematics thinking and learning. The second one concerns semiotics and intersubjectivity. The third one is about semiotics as the focus of innovative learning and teaching materials.

2.4.1 The Relationship Among Sign Systems and Translation

Duval's (2000, 2006) studies have been very important in showing the complexities behind the relationship between sign systems and the difficulties that the students encounter when faced with moving between semiotic registers. In this line of thought, an investigation into the meaning that students ascribe to their first algebraic formulas expressed through the standard algebraic symbolism suggested that their emerging meanings are deeply rooted in significations that come from natural language and perception. In the case of the translation of statements in natural language into the standard algebraic symbolism, Radford (2002b) noticed that the first algebraic statements are not only imbued with the meanings of colloquial language, but also colloquial language lends a specific mode of designation of objects that conflicts with the mode of designation of objects of algebraic symbolism. He discusses a mathematical activity that was based on the following short story: "Kelly has 2 more candies than Manuel. Josée has 5 more candies than Manuel. All together they have 37 candies." During the mathematical activity, in Problem 1, the students were invited to designate Manuel's number of candies by x, to elaborate a symbolic expression for Kelly and Josée, and, then, to write and solve an equation corresponding to the short story. In Problem 2, the students were invited to designate Kelly's number of candies by x while in Problem 3 the students were invited to designate Josée's number of candies by x. Radford suggests that one of the difficulties in dealing with problems involving comparative phrases like "Kelly has 2 more candies than Manuel" is being able to derive non-comparative, assertive phrases of the type: "A (or B) has C". If, say, Manuel has 4 candies, the assertive phrase would take the form «Kelly (Subject) has (Verb) 6 (Adjective) candies (Noun)». In the case of algebra, the adjective is not known (one does not know how many candies A has). As a result, the adjective has to be referred to in some way. In using a letter like 'x' (or another device) a new semiotic space is opened. In this space, the story problem has to be re-told, leading to what has been usually termed (although in a rather simplistic way) the 'translation' of the problem into an equation. Radford suggests the term *symbolic narrative*, arguing that what is 'translated' still tells us a story but in mathematical symbols. (Radford 2002b). He shows that some of the difficulties that the students have in operating with the symbols are precisely related to the requirement of producing a collapse in the original stated story. This he terms the *collapse of narratives*, adding

> The collection of similar terms means a rupture with their original meaning. All the efforts that were made at the level of the designation of objects to build the symbolic narrative have to be put into brackets. The whole symbolic narrative now has to collapse. There is no corresponding segment in the story-problem that could be correlated with the result of the collection [addition] of similar terms. (Radford 2002b, Vol. 4, p. 87)

The longitudinal investigation of several cohorts of students in pattern generalization point to a similar result: One of the crucial developmental steps in the students' algebraic thinking consists in moving from an indexical mode of designation to a symbolic one (see, e.g., Radford 2010b).

In a study of biological research from data collection to the published results, Latour (1993) shows how soil samples are translated into a sign system that is translated into other another sign system until, at the end, verbal statements about the biological system are made. In each case, the sign system consists of a material base with some structure. The relation between two sign systems is not inherent or natural but is established through work. Incidentally, a similar articulation was offered to understand how students relate a winch to pull up weights and mathematical (symbolic) structures, which have symbolic notations as their material

(Greeno 1988). To assist students in learning the relation between mathematical graphs and the physical phenomena they are investigating, some textbooks *layer* different sign systems with the apparent intention of offering students a way to link particular aspects of one system to a corresponding aspect in the other (Roth et al. 2005). Using an example from a Korean science textbook, a graph exhibiting Boyle's law relating the volume V and pressure of an ideal gas ($V \sim 1/p$) is overlaid by (a) the images of a glass beaker with different weights and (b) differently sized arrows (i.e., weight vectors) corresponding to the weights. The authors suggest that these complexes of sign systems may be difficult to unpack because relationships emerge only when students structure each layer in a particular way so that the desired relationships then can be constructed.

2.4.2   Semiotics and Intersubjectivity

A different take on intersubjectivity apparently arose from Vygotsky's "last, 'Spinozan' works [where] the idea of semiotic mediation is supplanted by the concept of the intersubjective speech field" (Mikhailov 2006, p. 35). Because children always already find themselves in an intersubjective speech field, the world and language are given to them as their own. As a result, there exists a "dynamic identity of intersubjectivity and intrasubjectivity" (p. 36). Subjectivity is a significant topic in its own right (e.g., Brown 2011), which can only be mentioned here.

2.4.3   Semiotics as the Focus of Innovative Learning and Teaching Materials

Digital mathematics textbooks, instructional materials integrating interactive diagrams, interactive visual examples and visual demonstration animations have constituted a privileged terrain of research in mathematics education. Semiotics helps to understand the challenges driven by these materials. Some important threads are as follows:

- Innovative visualization tools for teaching and learning;
- Design of activities and tasks that are based on interactive visual examples;
- Patterns of reading, using and solving with interactive linked multiple representations;
- Roles of diagrams, animations and video as instructional tools with new technologies.

One recent study exhibits the different types of work students need to accomplish to relate natural phenomena and different computer-based, dynamic sign systems—images and graphs—that are used to stand in for the former (Jornet and

Roth 2015). In each case, that is, in natural phenomenon and sign systems, structuring work is required to get from the material base to a structure. The structures of the different systems may then be related and compared, often leading to a revision in the structuring process, which enables new forms or relations between the different systems

Chapter 3
A Summary of Results

Within the constraints of this Topical Survey, we have of necessity concentrated on the many theoretical constructs that are relevant to semiotics in mathematics education. Applications are mentioned, but we have included more details of only one of the many empirical research studies that have been conducted. Interested readers may follow the rich empirical results contained in many of the references cited. Each of the topics in the following list of items surveyed has the potential to generate questions for further empirical research.

- Basic ideas and applications of theories of de Saussure, Peirce, Vygotsky, and other seminal thinkers;
- The roles of visualization and language in semiosis;
- Relevant theoretical notions such as objectification and communicative fields;
- Embodiment and gestures in semiosis;
- Semiotic chains, semiotic bundles, and semiotic nodes;
- Other dimensions: sign systems and translations among them; intersubjectivity; the creation of innovative learning and teaching materials.

© The Author(s) 2016
N. Presmeg, *Semiotics in Mathematics Education*,
ICME-13 Topical Surveys, DOI 10.1007/978-3-319-31370-2_3

References

Alrø, H., Ravn, O., & Valero, P. (2010). *Critical mathematics education: Past, present and future.* Rotterdam: Sense.

Anderson, M., Sáenz-Ludlow, A., Zellweger, S., & Cifarelli, V. (2003). *Educational perspectives on mathematics as semiosis: From thinking to interpreting to knowing.* Ottawa: Legas.

Arnheim, R. (1969). *Visual thinking.* The Regents of the University of California.

Arzarello, F. (2006). Semiosis as a multimodal process. *Revista Latinoamericana De Investigación En Matemática Educativa, Special Issue on Semiotics, Culture, and Mathematical Thinking (Guest Editors: L. Radford & B. D'Amore)* (pp. 267–299).

Arzarello, F., Paola, D., Robutti, O., & Sabena, C. (2009). Gestures as semiotic resources in the mathematics classroom. *Educational Studies in Mathematics, 70*(2), 97–109.

Bachmann-Medick, D. (2009). *Cultural turns: Neuorientierungen in den Kulturwissenschaften.* Reinbek bei Hamburg, Germany: Rowohlt.

Bakhtin, M. M. (1981). *The dialogic imagination.* Austin, TX: University of Texas Press.

Bakhtin, M. M. (1984). *Problems of Dostoevsky's poetics.* Austin, TX: University of Texas Press.

Bartolini Bussi, M., & Mariotti, M. A. (2008). Semiotic mediation in the mathematics classroom: Artefacts and signs after a vygotskian perspective. In L. English (Ed.), *Handbook of international research in mathematics education* (2nd ed., pp. 746–783). New York: Routledge, Taylor and Francis.

Barwell, R. (2015). Formal and informal mathematical discourses: Bakhtin and Vygotsky dialogue and dialectic. *Educational Studies in Mathematics.* doi:10.1007/s10649-015-9641-z.

Bautista, A., & Roth, W.-M. (2012). Conceptualizing sound as a form of incarnate mathematical consciousness. *Educational Studies in Mathematics, 79*(1), 41–59.

Bender, J. B., & Marrinan, M. (2010). *The culture of diagram.* Stanford: Stanford University Press.

Boehm, G. (1994). *Was ist ein Bild?.* München, Germany: Fink.

Braathe, H. J., & Solomon, Y. (2015). Choosing mathematics: The narrative of the self as a site of agency. *Educational Studies in Mathematics, 89*, 151–166.

Brousseau, G. (1997). *Theory of didactical situations in mathematics.* Dordrecht: Kluwer.

Brown, T. (2011). *Mathematics education and subjectivity.* Dordrecht: Springer.

Charleton, T. L. (1996). *An elementary Latin dictionary.* Oxford, England: Oxford University Press.

Colapietro, V. M. (1993). *Glossary of semiotics.* New York, NY: Paragon House.

de Freitas, E., & Sinclair, N. (2013). New materialist ontologies in mathematics education: The body in/of mathematics. *Educational Studies in Mathematics, 83*, 453–470.

de Saussure, F. (1959). *Course in general linguistics.* New York, NY: McGraw-Hill.

de Waal, C. (2013). *Peirce: A guide to the perplexed.* New York: Bloomsbury.

Donham, D. L. (1999). *History, power, ideology: Central issues in marxism and anthropology.* Berkeley: University of California Press.

Dubinsky, E. (2002). Reflective abstraction in advanced mathematical thinking. In D. Tall (Ed.), *advanced mathematical thinking* (pp. 95–123). New York: Kluwer.

Dubinsky, E., & McDonald, M. (2001). APOS: A constructivist theory of learning in undergraduate mathematics education research. In *The teaching and learning of mathematics at university level: An ICMI study* (pp. 275–282). Dordret: Kluwer.

Duval, R. (1995). *Sémoisis et pensée humaine [Semiosis and human thinking]*. Bern: Lang.

Duval, R. (2000). Basic issues for research in mathematics education. In Nakahara & M. Koyama (Eds.), *Proceedings of the 24th conference of the International Group for the Psychology of Mathematics Education (PME-24)* (Vol. 1, pp. 55–69). Japan: Hiroshima University.

Duval, R. (2006). A cognitive analysis of problems of comprehension in a learning of mathematics. *Educational Studies in Mathematics, 61*, 103–131.

Eco, U. (1986). *Semiotics and the philosophy of language*. Bloomington, IN: Indiana University Press.

Eco, U. (1988). *Le signe [The sign]*. Bruxelles: Éditions Labor.

El'konin, B. D. (1994). *Vvedenie v psixologiju razvitija: B tradicii kul'turno-istoričeskoj teorii L. S. Vygotskogo* [Introduction to the psychology of development: In the tradition of the cultural-historical theory of L. S. Vygotsky]. Moscow, Russia: Trivola.

Ernest, P. (2006). A semiotic perspective of mathematical activity: The case of number. *Educational Studies in Mathematics, 61*, 67–101.

Finn, J. (Ed.). (2012a). *Visual communication and culture: Images in action*. Ontario: Oxford University Press.

Finn, J. (2012b). Powell's point: Denial and deception at the UN. In J. Finn (Ed.), *Visual communication and the culture: Images in action* (pp. 200–217). Ontario: Oxford University Press.

Font, V., Godino, J., & Gallardo, J. (2013). The emergence of mathematical objects from mathematical practices. *Educational Studies in Mathematics, 82*(1), 97–124.

Fried, M. N. (2007). Didactics and history of mathematics: Knowledge and self-knowledge. *Educational Studies in Mathematics, 66*(2), 203–223.

Fried, M. N. (2008). History of mathematics in mathematics education: A Saussurean perspective. *The Montana Mathematics Enthusiast, 5*, 185–198.

Goldin, G. A., & Janvier, C. (1998). Representations and the psychology of mathematics education. *The Journal of Mathematical Behavior, 17.1 & 17.2*.

González Rey, F. (2009). Historical relevance of Vygotsky's work: Its significance for a new approach to the problem of subjectivity in psychology. *Outlines, 1*, 59–73.

González Rey, F. (2011a). *El pensamiento de Vigotsky. Contradicciones, desdoblamientos y desarrollo [Vygotsky's thought. Contradictions, splittings, and development]*. Mexico: Trillas.

González Rey, F. (2011b). A re-examination of defining moments in vygotsky's work and their implications for his continuing legacy. *Mind, Culture, and Activity, 18*, 257–275.

Greeno, J. G. (1989). Situations, mental models, and generative knowledge. In D. Klahr & K. Kotovsky (Eds.), *Complex information processing: The impact of Herbert A. Simon* (pp. 285–318). Hillsdale, NJ: Lawrence Erlbaum Associates.

Halliday, M. (1978). *Language as social semiotic*. London: Arnold.

Hasan, R. (2002). Semiotic mediation, language and society: Three exotropic theories–Vygotsky, Hallyday and Bernstein. Retrieved June 20, 2007, from http://www.education.miami.edu/blantonw/mainsite/Componentsfromclmer/Component13/Mediation/SemioticMediation.htm.

Heidegger, M. (1977). *Sein und Zeit [Being and time]*. Tübingen: Max Niemeyer.

Henry, M. (2000). *Incarnation: Une philosophie de la chair [Incarnation: A philosophy of the flesh]*. Paris: Éditions du Seuil.

Heßler, M., & Mersch, D. (Eds.). (2009). *Logik des Bildlichen: Zur Kritik der ikonischen Vernunft*. Bielefeld, Germany: Transcript Verlag.

Holzkamp, K. (1983). *Grundlegung der Psychologie [Founding (Foundation of) psychology]*. Frankfurt: Campus.

Husserl, E. (1939). Die Frage nach dem Ursprung der Geometrie als intentional-historisches Problem [The question about the origin of geometry as intentional-historical problem]. *Revue internationale de philosophie, 1*, 203–225.

Johnson, M. (1987). *The body in the mind: The bodily basis of imagination, reason, and meaning.* Chicago: Chicago University Press.

Jornet, A., & Roth, W.-M. (2015). The joint work of connecting multiple (re)presentations in science classrooms. *Science Education, 99*, 378–403.

Kazak, S., Wegerif, R., & Fujita, T. (2015). The importance of dialogic processes to conceptual development in mathematics. *Educational Studies in Mathematics, 90*, 105–120.

Knijnik, G. (2012). Differentially positioned language games: Ethnomathematics from a philosophical perspective. *Educational Studies in Mathematics, 80*, 87–100.

Lacan, J. (1966). *Écrits.* Paris: Éditions du Seuil.

Lakoff, G. (1987). *Women, fire, and dangerous things: What categories reveal about the mind.* Chicago: University of Chicago Press.

Lakoff, G., & Núñez, R. (2000). *Where mathematics comes from: How the embodied mind brings mathematics into being.* New York, NY: Basic Books.

Latour, B. (1993). *La clef de Berlin et autres leçons d'un amateur de sciences.* Paris: Éditions la Découverte.

Leontiev (or Leont'ev), A. N. (1968). El hombre y la cultura [Man and culture]. In *El hombre y la cultura: Problemas teóricos sobre educación* (pp. 9–48). Mexico: Editorial Grijalbo.

Leontyev, A. N. (1981). *Problems of the development of the mind.* Moscow, USSR: Progress Publishers.

Leont'ev (or Leontyev), A. N. (2009). *Activity and consciousness.* Pacifica, CA: MIA. Retrieved August 29, 2009, from http://www.marxists.org/archive/leontev/works/activity-consciousness. pdf.

Livingston, E. (1986). *The ethnomethodological foundations of mathematics.* London, UK: Routledge & Kegan Paul.

Maine de Biran, P. (1841). *Œuvres philosophiques tome premier: Influence de l'habitude sur la faculté de penser* [Philosophical words vol. 1: Influence of habitude on the capacity to think]. Paris: Librairie de Ladrange.

Mariotti, M. A. & Bartolini Bussi, M. G. (1998). From drawing to construction: Teachers' mediation within the Cabri environment. In A. Olivier & K. Newstead (Eds.), *Proceedings of the 22nd PME International Conference* (Vol. 1, pp. 180–195).

Mikhailov, F. T. (2001). The "Other Within" for the psychologist. *Journal of Russian and East European Psychology, 39*(1), 6–31.

Mikhailov, F. T. (2004). Object-oriented activity—Whose? *Journal of Russian and East European Psychology, 42*(3), 6–34.

Mikhailov, F. T. (2006). Problems of the method of cultural-historical psychology. *Journal of Russian and East European Psychology, 44*(1), 21–54.

Mirzoeff, N. (Ed.). (2002). *The visual culture reader.* New York: Routledge.

Mitchell, W. J. T. (1987). *Iconology: image, text, ideology.* Chicago: University of Chicago Press.

Morgan, C. (2006). What does social semiotics have to offer mathematics education research? *Educational Studies in Mathematics, 61*, 219–245.

Morgan, C. (2009). Understanding practices in mathematics education: Structure and text. In M. Tzekaki, M. Kaldrimidou, & H. Sakonidis (Eds.), *Proceedings of the 33rd conference of the International Group for the Psychology of Mathematics Education* (pp. 49–66). Greece: Thessaloniki.

Morgan, C. (2012). Studying discourse implies studying equity. In B. Herbel-Eisenmann, J. Choppin, D. Wagner, & D. Pimm (Eds.), *Equity in discourse for mathematics education* (pp. 181–192). New York: Springer.

Nöth, W. (1990). *Handbook of semiotics.* Bloomington, IN: Indiana University Press.

Núñez, R. E. (2009). Gesture, inscriptions, and abstraction: The embodied nature of mathematics or why mathematics education shouldn't leave the math untouched. In W.-M. Roth (Ed.), *Mathematical representation at the interface of body and culture* (pp. 309–328). Charlotte, NC: Information Age Publishing.

Otte, M. (2006). Mathematics epistemology from a Peircean semiotic point of view. *Educational Studies in Mathematics, 61*, 11–38.

Pais, A. (2013). *A Critique of Ideology on the Issue of Transfer, 84*, 15–34.

Peirce, C. S. (1931–1958). *Collected papers* (CP, Vols. 1–8). Cambridge: Harvard University Press.

Peirce, C. S. (1992). *The essential Peirce.* In N. Houser & C. Kloesel (Ed.) (Vol. 1). Bloomington: Indiana University Press.

Peirce, C. S. (1998). *The essential Peirce.* In The Peirce Edition Project (Ed.) (Vol. 2). Bloomington, Indiana: Indiana University Press.

Popkewitz, T. (2004). School subjects, the politics of knowledge, and the projects of intellectuals in change. In P. Valero & R. Zevenbergen (Eds.), *Researching the socio-political dimensions of mathematics education* (pp. 251–267). Boston: Kluwer.

Presmeg, N. C. (1985). *Visually mediated processes in high school mathematics: A classroom investigation.* Unpublished Ph.D. dissertation, University of Cambridge.

Presmeg, N. C. (1992). Prototypes, metaphors, metonymies, and imaginative rationality in high school mathematics. *Educational Studies in Mathematics, 23*, 595–610.

Presmeg, N. C. (1997). A semiotic framework for linking cultural practice and classroom mathematics. In J. Dossey, J.Swafford, M. Parmantie, & A. Dossey (Eds.), *Proceedings of the 19th Annual Meeting of the North American Chapter of the International Group for the Psychology of Mathematics Education* (Vol. 1, pp. 151–156). Columbus, Ohio.

Presmeg, N. C. (1998). Ethnomathematics in teacher education. *Journal of Mathematics Teacher Education, 1*(3), 317–339.

Presmeg, N. C. (2003). Ancient areas: A retrospective analysis of early history of geometry in light of Peirce's "commens". Paper presented in Discussion Group 7, *Semiotic and socio-cultural evolution of mathematical concepts*, 27th Annual Meeting of the International Group for the Psychology of Mathematics Education, Honolulu, Hawaii, July 13–18, 2003. Subsequently published in the *Journal of the Svensk Förening för MatematikDidaktisk Forskning (MaDiF)* (Vol. 8, pp. 24–34), December, 2003.

Presmeg, N. C. (2006a). A semiotic view of the role of imagery and inscriptions in mathematics teaching and learning. In J. Novotná, H. Moraová, M. Krátká, & N. Stehlikóvá (Eds.), *Proceedings of the 30th Conference of the International Group for the Psychology of Mathematics Education* (Vol. 1, pp. 19–34). Prague: PME.

Presmeg, N. C. (2006b). Semiotics and the "connections" standard: Significance of semiotics for teachers of mathematics. *Educational Studies in Mathematics, 61*, 163–182.

Presmeg, N. C. (2014). Contemplating visualization as an epistemological learning tool in mathematics. *ZDM—The International Journal on Mathematics Education, 46*(1), 151–157. Commentary paper, in the issue on *Visualization as an epistemological learning tool*, guest-edited by F. D. Rivera, H. Steinbring, & A. Arcavi.

Radford, L. (2000). Signs and meanings in students' emergent algebraic thinking: a semiotic analysis. *Educational Studies in Mathematics, 42*, 237–268.

Radford, L. (2002a). The seen, the spoken, and the written: A semiotic approach to the problem of objectification of mathematical knowledge. *For the Learning of Mathematics, 22*, 14–23.

Radford, L. (2002b). On heroes and the collapse of narratives. A contribution to the study of symbolic thinking. In A. D. Cockburn & E. Nardi (Eds.), *Proceedings of the 26th Conference of the International Group for the Psychology of Mathematics Education, PME 26* (Vol. 4, pp. 81–88). UK: University of East Anglia.

Radford, L. (2003). Gestures, speech and the sprouting of signs. *Mathematical Thinking and Learning, 5*(1), 37–70.

Radford, L. (2006a). How to look at the general through the particular: Berkeley and Kant on symbolizing mathematical generality. In S. Sbaragli (Ed.), *La matematica e la sua didattica* (pp. 245–248). Roma: Carocci Faber.

Radford, L. (2006b). Elementos de una teoría cultural de la objetivación [elements of a cultural theory of objectification]. *Revista Latinoamericana de Investigación en Matemática Educativa, Special Issue on Semiotics, Culture and Mathematical Thinking* (pp. 103–129) (English version available at: http://www.luisradford.ca).

Radford, L. (2008a). Diagrammatic thinking: Notes on Peirce's semiotics and epistemology. *PNA, 3*(1), 1–18.

Radford, L. (2008b). The ethics of being and knowing: Towards a cultural theory of learning. In L. Radford, G. Schubring, & F. Seeger (Eds.), *Semiotics in mathematics education: Epistemology, history, classroom, and culture* (pp. 215–234). Rotterdam: Sense Publishers.

Radford, L. (2009). Why do gestures matter? Sensuous cognition and the palpability of mathematical meanings. *Educational Studies in Mathematics, 70*(2), 111–126.

Radford, L. (2010a). The eye as a theoretician: Seeing structures in generalizing activities. *For the Learning of Mathematics, 30*(2), 2–7.

Radford, L. (2010b). Algebraic thinking from a cultural semiotic perspective. *Research in Mathematics Education, 12*(1), 1–19.

Radford, L. (2012). On the cognitive, epistemic, and ontological roles of artifacts. In G. Gueudet, B. Pepin, & L. Trouche (Eds.), *From text to 'lived' resources mathematics curriculum materials and teacher development* (pp. 282–288). New York: Springer.

Radford, L. (2013a). On semiotics and education. *Éducation & Didactique, 7*(1), 185–204.

Radford, L. (2013b). Three key concepts of the theory of objectification: Knowledge, knowing, and learning. *Journal of Research in Mathematics Education, 2*(1), 7–44.

Radford, L. (2014a). Towards an embodied, cultural, and material conception of mathematics cognition. *ZDM—The International Journal on Mathematics Education, 46*, 349–361.

Radford, L. (2014b). On the role of representations and artefacts in knowing and learning. *Educational Studies in Mathematics, 85*, 405–422.

Radford, L. (2014c). On teachers *and* students: An ethical cultural-historical perspective. In P. Liljedahl, C. Nicol, S. Oesterle, & D. Allan (Eds.), *Proceedings of the joint meeting of PME 38 and PME-NA 36* (Vol. 1, pp. 1–20). Vancouver: PME.

Radford, L. (2015a). Methodological aspects of the theory of objectification. *Perspectivas Da Educação Matemática, 8*(18), 547–567.

Radford, L. (2015b). The epistemological foundations of the theory of objectification. *Isonomia* (http://isonomia.uniurb.it/epistemologica) (pp. 127–149).

Radford, L., Arzarello, F., Edwards, L., & Sabena, C. (in press). The multimodal material mind: Embodiment in mathematics education. In J. Cai (Ed.), *First compendium for research in mathematics education*. Reston, VA: National Council of Teachers of Mathematics.

Radford, L., Demers, S., Guzmán, J., & Cerulli, M. (2003). Calculators, graphs, gestures, and the production of meaning. In P. Pateman, B. Dougherty, & J. Zilliox (Eds.), *Proceedings of the 27 conference of the International Group for the Psychology of Mathematics Education (PME27 - PMENA25)* (Vol. 4, pp. 55–62). University of Hawaii.

Radford, L., & Empey, H. (2007). Culture, knowledge and the self: Mathematics and the formation of nw social sensibilities in the renaissance and medieval islam. *Revista Brasileira De História Da Matemática. Festschrift Ubiratan D'Ambrosio* (pp. 231–254).

Radford, L., & Sabena, C. (2015). The question of method in a Vygotskian semiotic approach. In A. Bikner-Ahsbahs, C. Knipping, & N. Presmeg (Eds.), *Approaches to qualitative research in mathematics education* (pp. 157–182). New York: Springer.

Radford, L., Schubring, G., & Seeger, F. (2008). *Semiotics in mathematics education: Epistemology, history, classroom, and culture*. Rotterdam: Sense Publishers.

Radford, L., Schubring, G., & Seeger, F. (2011). Signifying and meaning-making in mathematics thinking, teaching and learning: Semiotic perspectives. *Educational Studies in Mathematics, 77*(2–3), 149–397.

Roth, W.-M. (2003a). Competent workplace mathematics: How signs become transparent in use. *International Journal of Computers for Mathematical Learning, 8*, 161–189.

Roth, W.-M. (2003b). From epistemic (ergotic) actions to scientific discourse: Do gestures obtain a bridging function? *Pragmatics & Cognition, 11*, 139–168.

Roth, W.-M. (2006). A dialectical materialist reading of the sign. *Semiotica, 160*, 141–171.

Roth, W.-M. (2008). The dawning of signs in graph interpretation. In L. Radford, G. Schubring, & F. Seeger (Eds.), *Semiotics in mathematics education* (pp. 83–102). Rotterdam: Sense Publishers.

Roth, W.-M. (2010). Incarnation: Radicalizing the embodiment of mathematics. *For the Learning of Mathematics, 30*(2), 8–17.

Roth, W.-M. (2011). *Geometry as objective science in elementary classrooms: Mathematics in the flesh*. New York: Routledge.

Roth, W.-M. (2012). Tracking the origins of signs in mathematical activity: A material phenomenological approach. In M. Bockarova, M. Danesi, & R. Núñez (Eds.), *Cognitive science and interdisciplinary approaches to mathematical cognition* (pp. 209–247). Munich, Germany: LINCOM EUROPA.

Roth, W.-M. (2013). An integrated theory of thinking and speaking that draws on Vygotsky and Bakhtin/Vološinov. *Dialogical Pedagogy, 1*, 32–53.

Roth, W.-M. (2014). Science language Wanted Alive: Through the dialectical/dialogical lens of Vygotsky and the Bakhtin circle. *Journal of Research in Science Teaching, 51*, 1049–1083.

Roth, W.-M. (2015). The emergence of signs in hands-on science. In P. Trifonas (Ed.), *International handbook of semiotics* (pp. 1271–1289). Dordrecht: Springer.

Roth, W.-M. (2016a). Birth of signs: From triangular semiotics to communicative fields. Paper to be presented in Topic Study Group 54, *Semiotics in mathematics education*, International Congress on Mathematical Education, Hamburg, Germany, July 24–31, 2016.

Roth, W.-M. (2016b). *Concrete human psychology*. New York: Routledge.

Roth, W.-M. (in press). Cultural concepts concretely. In E. de Freitas, N. Sinclair, & A. Coles (Eds.), *What is a mathematical concept*. Cambridge: Cambridge University Press.

Roth, W.-M., & Bowen, G. M. (2001). Professionals read graphs: A semiotic analysis. *Journal for Research in Mathematics Education, 32*, 159–194.

Roth, W.-M., & Bowen, G. M. (2003). When are graphs ten thousand words worth? An expert/expert study. *Cognition and Instruction, 21*, 429–473.

Roth, W.-M., Bowen, G. M., & Masciotra, D. (2002). From thing to sign and 'natural object': Toward a genetic phenomenology of graph interpretation. *Science, Technology, & Human Values, 27*, 327–356.

Roth, W.-M., & Lawless, D. (2002). Signs, deixis, and the emergence of scientific explanations. *Semiotica, 138*, 95–130.

Roth, W.-M., Pozzer-Ardenghi, L., & Han, J. (2005). *Critical graphicacy: Understanding visual representation practices in school science*. Dordrecht: Springer-Kluwer.

Roth, W.-M., & Radford, L. (2011). *A cultural historical perspective on teaching and learning*. Rotterdam: Sense Publishers.

Sáenz-Ludlow, A., & Kadunz, G. (2016). *Semiotics as a tool for learning mathematics: How to describe the construction, visualisation, and communication of mathematics concepts*. Rotterdam: Sense Publishers.

Sáenz-Ludlow, A., & Presmeg, N. (2006). Semiotic perspectives in mathematics education. *Educational Studies in Mathematics. Special Issue, 61*(1–2).

Sfard, A. (2001). There is more to discourse than meets the ears: Looking at thinking as communicating to learn more about mathematical learning. *Educational Studies in Mathematics, 46*, 13–57.

Sfard, A. (2008). *Thinking as communicating*. Cambridge: Cambridge University Press.

Sfard, A. (2010). The challenges of researching discursive practice in classrooms. In M. Pinto & T. Kawasaki (Eds.), *Proceedings of the 34th conference of the international group for the psychology of mathematics education* (Vol. 1, pp. 217–221). Belo Horizonte, Brazil: PME.

Sheets-Johnstone, M. (2009). *The corporeal turn: An interdisciplinary reader*. Exeter, UK: Imprint Academic.

Skovsmose, O. (2008). Critical mathematics education for the future. In M. Niss (Ed.), *ICME-10 proceedings*. (Retrieved on December 20 2010 from http://www.icme10.dk/proceedings/pages/regular_pdf/RL_Ole_Skovsmose.pdf). Denmark: IMFUFA, Department of Science, Systems and Models, Roskilde University.

Solomon, Y. (2012). Finding a voice? Narrating the female self in mathematics. *Educational Studies in Mathematics, 80*, 171–183.

Styhre, A. (2010). *Visual culture in organizations: Theory and cases* (Vol. 9). New York: Routledge.

Tall, D. (2004). Building theories: The three worlds of mathematics. *For the learning of mathematics* (pp. 29–32).

Tall, D. (2008). The transition to formal thinking in mathematics. *Mathematics Education Research Journal, 20*(2), 5–24.

Tall, D. (2013). *How humans learn to think mathematically*. Cambridge: Cambridge University Press.

Vergnaud, G. (1985). Concepts et schèmes dans la théorie opératoire de la representation [Concepts and schemas in the operatory theory of representation]. *Psychologie Française, 30*(3–4), 245–252.

Vilela, D. S. (2010). Discussing a philosophical background for the ethnomathematical program. *Educational Studies in Mathematics, 75*, 345–358.

Vološinov, V. N. (1930). *Marksizm i folosofija jazyka: osnovye problemy sociologičeskogo metoda b nauke o jazyke [Marxism and the philosophy of language: Main problems of the sociological method in linguistics]*. Leningrad, USSR: Priboj.

Voloshinov [or Vološinov], V. N. (1973). *Marxism and the philosophy of language*. New York: Seminar Press.

Vygotsky, L. S. (1929). The problem of the cultural development of the child. *Journal of Genetic Psychology, 36*, 415–434.

Vygotsky, L. S. (1971). *The psychology of art*. Cambridge London: MIT Press (First published in 1925).

Vygotsky, L. S. (1978). *Mind in society*. Cambridge, MA: Harvard University Press.

Vygotsky, L. (1979). Consciousness as a problem in the psychology of behavior. *Soviet Psychology, 17*(4), 3–35.

Vygotsky, L. S. (1987). *The collected works of L. S. Vygotsky, vol. 1: Problems of general psychology*. New York, NY: Springer.

Vygotsky, L. S. (1989). Concrete human psychology. *Soviet Psychology, 27*(2), 53–77.

Vygotsky, L. S. (1993). *Collected works* (Vol. 2). New York: Plenum.

Vygotsky, L. (1997). *Collected works* (Vol. 3). New York: Plenum.

Vygotsky, L. S. (1999). *Collected works* (Vol. 6). New York: Plenum.

Walkerdine, V. (1988). *The mastery of reason: Cognitive developments and the production of rationality*. New York: Routledge.

Wartofsky, M. (1968). *Conceptual foundations of scientific thought*. New York: Macmillan.

Whitson, J. A. (1994). Elements of a semiotic framework for understanding situated and conceptual learning. In D. Kirshner (Ed.), *Proceedings of the 16th Annual Meeting of the North American Chapter of the International Group for the Psychology of Mathematics Education* (Vol. 1, pp. 35–50). Baton Rouge, LA.

Whitson, J. A. (1997). Cognition as a semiosic process: From situated mediation to critical reflective transcendence. In In D. Kirshner and J. A. Whitson (Eds.), *Situated cognition: Social, semiotic, and psychological perspectives*. Mahwah, NJ: Lawrence Erlbaum Associates.

Wittgenstein, L. (1997). *Philosophical Investigations/Philosophische Untersuchungen* (2nd ed.). Oxford, UK: Blackwell. (First published in 1953).

Yasnitsky, A. (2011). The Vygotsky that we (do not) know: Vygotsky's main works and the chronology of their composition. *PsyAnima, Dubna Psychological Journal, 4*, 52–70. Retrieved from www.psyanima.ru.

Further Reading: Recommended General Literature from the Reference List

Radford, L. (2014). On the role of representations and artefacts in knowing and learning. *Educational Studies in Mathematics, 85*, 405–422.

Radford, L., Schubring, G., & Seeger, F. (2008). *Semiotics in mathematics education: Epistemology, history, classroom, and culture*. Rotterdam: Sense Publishers.

Roth, W.-M., & Radford, L. (2011). *A cultural historical perspective on teaching and learning*. Rotterdam: Sense Publishers.

Sáenz-Ludlow, A., & Kadunz, G. (2016). *Semiotics as a tool for learning mathematics: How to describe the construction, visualisation, and communication of mathematics concepts*. Rotterdam: Sense Publishers.

Sáenz-Ludlow, A., & Presmeg, N. (2006). Semiotic perspectives in mathematics education. *Educational Studies in Mathematics. Special Issue, 61*(1–2).